CHEERS FOR CAREERS!

I WANT TO BE AN ELECTRICIAN

by Julie Murray

Cody Koala
An Imprint of Pop!
popbooksonline.com

Hello! My name is Cody Koala

This book is filled with videos, puzzles, games, and more! Scan the QR codes* while you read, or visit the website below to make this book pop.

popbooksonline.com/electrician

*Scanning QR codes requires a web-enabled smart device with a QR code reader app and a camera.

abdobooks.com
Published by Pop!, a division of ABDO, PO Box 398166, Minneapolis, Minnesota 55439.

Printed in the United States of America, North Mankato, Minnesota.
052024
082024
THIS BOOK CONTAINS RECYCLED MATERIALS

Cover Photo: Shutterstock Images
Interior Photos: Getty Images, Shutterstock Images
Editor: Elizabeth Andrews and Grace Hansen
Series Designer: Colleen McLaren

Library of Congress Control Number: 2023947424

Publisher's Cataloging-in-Publication Data
Names: Murray, Julie, author.
Title: I want to be an electrician / by Julie Murray
Description: Minneapolis, Minnesota : Pop!, 2025 | Series: Cheers for careers! | Includes online resources and index
Identifiers: ISBN 9781098246099 (lib. bdg.) | ISBN 9781098246655 (ebook)
Subjects: LCSH: Electricians--Juvenile literature. | Electrical engineering--Juvenile literature. | Electric power supply--Juvenile literature. | Servicing trades--Juvenile literature. | Occupations--Juvenile literature.
Classification: DDC 621.38--dc23

Table of Contents

Chapter 1

Electrical Power

Electricians **install** and repair electrical systems. They also work to keep electrical systems running well in buildings and homes.

Watch a video here!

Chapter 2

Skills Needed

Electricians need to have math skills. They must be able to read plans called blueprints. They also need to be able to work with their hands.

Electricians work with many electrical systems and tools.

Learn more here!

Electricians also need to have good problem-solving skills. This means that they must be able to find the problem and come up with a plan to fix it.

Chapter 3

Becoming an Electrician

A person needs hands-on experience to become an electrician. One can get this through an **apprenticeship**. Classes are often taken during this time too.

Explore links here!

Next, a person must pass an exam. They can now be called a **journeyman**. A person must work at least two years at this level.

Journeymen can work on their own without supervision.

To become a master electrician, a journeyman must work around 4,000 hours. Electricians strengthen their skills during this time. Finally, another exam must be taken and passed.

wire stripper

side cutter

needle nose pliers

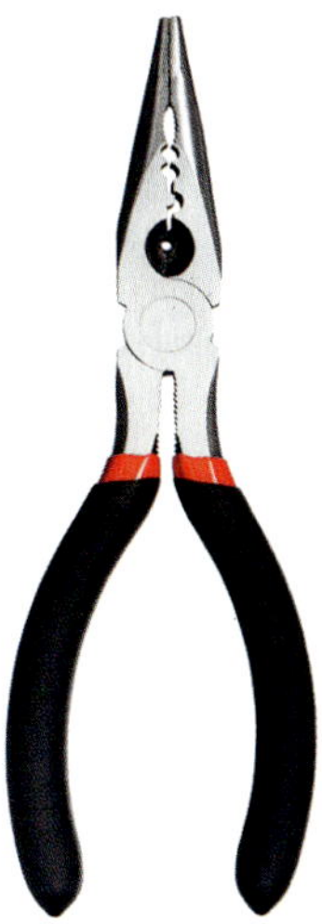

electrical tape

screwdrivers

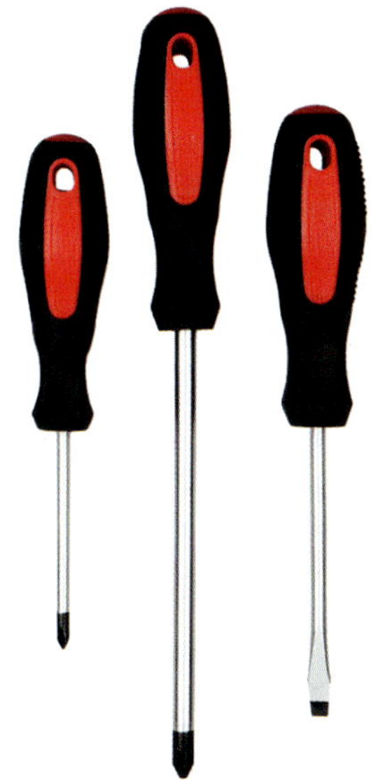

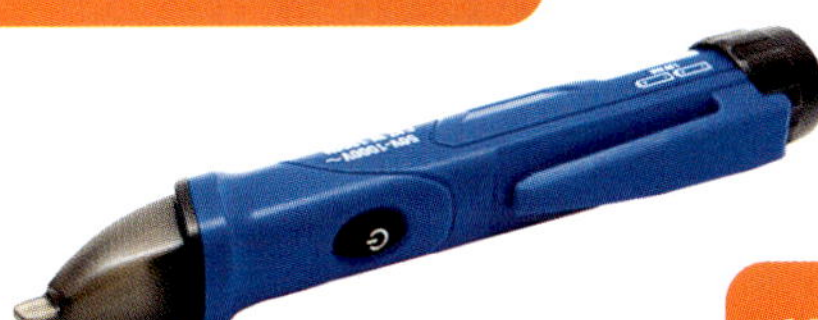

voltage tester

linesman pliers

wall chaser

multimeter

Chapter 4

Different Jobs

Electricians work in a variety of places. Many work in homes and office buildings. They work on **appliances**, lighting, air conditioning units, and much more.

Complete an activity here!

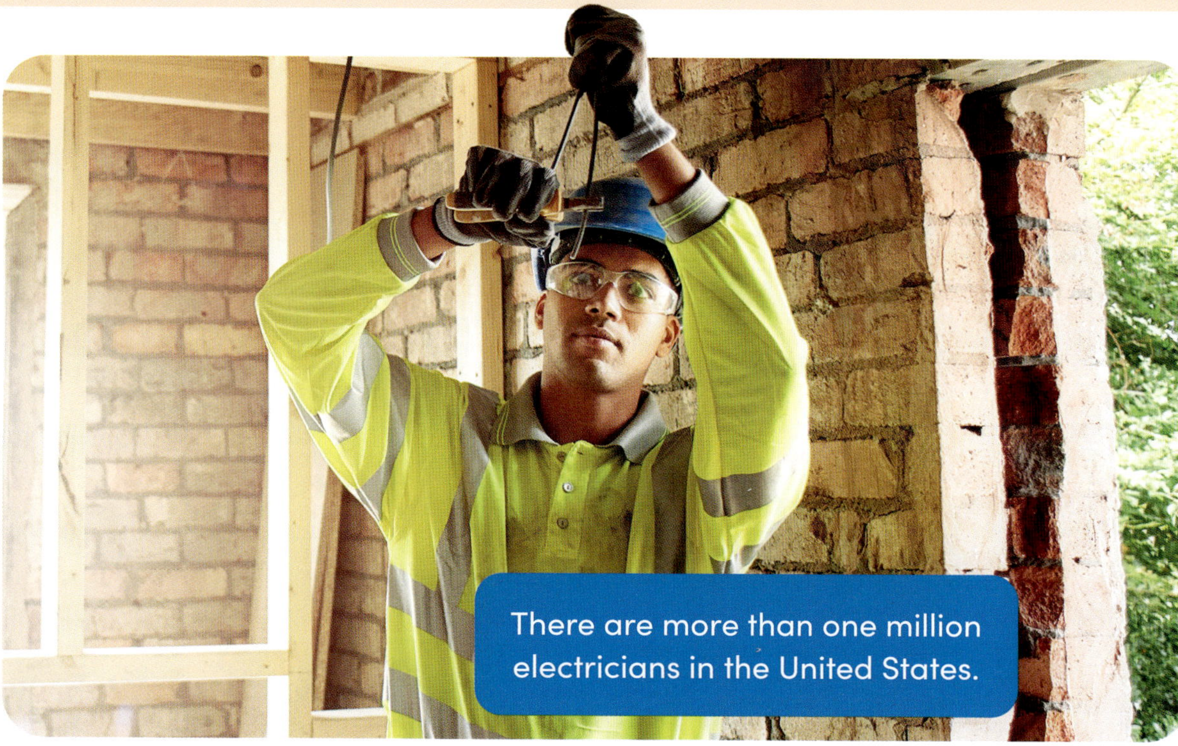

There are more than one million electricians in the United States.

Some electricians work in **construction**. They wire new buildings with electricity. Some work outside for

power companies. They are trained in **high-voltage** electrical work.

There is high demand for electricians. Many things we use each day require electricity. Electricians keep lights, heaters, and appliances running well. Being an electrician is a rewarding career!

Making Connections

Text-to-Self

Have you ever had the electricity go out at home or school? What did you have to do differently without electricity?

Text-to-Text

Have you read a book about something that uses electricity? What did you learn from that book?

Text-to-World

What are some things that run on electricity? Think about your home, school, stores, and anywhere else you go.

Glossary

appliance – a device used for a particular purpose. Stoves, refrigerators, and toasters are some appliances used in the home. Most appliances use electricity.

apprenticeship – an arrangement in which someone learns a trade under an expert.

construction – an area of work involving building, repairing, renovating, and maintaining structures.

high-voltage – marked by great energy.

install – to put into position and make ready for use.

journeyman – a worker who has completed an apprenticeship and works for another person.

Index

Online Resources

popbooksonline.com

Thanks for reading this Cody Koala book!

This book is filled with videos, puzzles, games, and more! Scan the QR codes* while you read, or visit the website below to make this book pop.

popbooksonline.com/electrician

*Scanning QR codes requires a web-enabled smart device with a QR code reader app and a camera.